THE CALCULATOR COMIC BOOK!

- Solve Complex Equations
- Learn How to Master Your Graphing Calculator
- Practice for National Exams
- Easy-to-Follow Examples

THE CALCULATOR COMIC BOOK!

DAVE PROCHNOW

First Edition

THE CALCULATOR COMIC BOOK!, FIRST EDITION

Copyright © 2016 by Dave Prochnow. All rights reserved. Printed in the United States of America. Except as permitted under the United States Copyright Act of 1976, no part of this publication may be reproduced or distributed in any form or by any means, or stored in a database or retrieval system, without the prior written permission of the author.

Camellia Pointe Publishing

Information contained in this work has been obtained by Camellia Pointe Publishing from sources believed to be reliable. However, neither Camellia Pointe Publishing nor its authors guarantee the accuracy or completeness of any information published herein, and neither Camellia Pointe Publishing nor its authors shall be responsible for any errors, omissions, or damages arising out of use of this information. This work is published with the understanding that Camellia Pointe Publishing and its authors are supplying information but are not attempting to render engineering or other professional services. If such services are required, the assistance of an appropriate professional should be sought.

Casio, Hewlett-Packard (HP), Sharp, and Texas Instruments (TI) are the trademarks of their respective owners with no affiliation with Camellia Pointe Publishing and/or Dave Prochnow.

ON MARCH 18, 1964, SHARP CORPORATION INTRODUCED THE WORLD'S FIRST ALL-TRANSISTOR DESKTOP CALCULATOR, THE CS-10A. ORIGINALLY PRICED AT 500,000 YEN.

Foreword

I'm sitting here in my dark, dank, dungeon, err, posh, spacious office overlooking the beautiful University of Transylvania campus, sharing my desktop with a calculator classic. This Hewlett-Packard HP-20S scientific calculator has been my constant computing companion since 1988--companion, nee, champion. Shortly after the acquisition of my beloved HP, I entered "57,005" [DEC], and converted it into hexadecimal base [HEX]. From that point onward, I was hooked on calculators.

Today's realistic math-like display graphing calculators are handheld marvels of technology. Stumped by an algebraic expression; just key it into one of these modern calculating wunderbars, EXACTLY as the formula appears in your textbook, and you'll have a solution, plus graph, faster than a bolt of lightning energizing an electrode. I should know, right?

Sure, devices like smartphones can mimic much of the computational prowess found in a calculator, but the programmable, battery-sipping, number-crunching strengths found in the typical handheld calculator impress more than just Professor Stein. Both educators and scholastic test review boards agree-- calculators, "YES," smartphones, "NO." Therefore, arm yourself with the power to pass any test, at any time.

My HP-20S was a difficult purchase decision back in 1988. At the big box stores today, however, you can buy math-like "textbook" display calculators for less than $10! Each!! That's an incredible bargain. Think of it, you can equip an entire math classroom of students for less than the retail price of my HP calculator. Even the more powerful graphing calculators can be obtained for about $100. At these remarkable discount prices, EVERYONE should load up on calculators as a specialized tool supplement to the ubiquitous smartphone.

Luckily, a calculator isn't a one-trick pony. Almost all of the scientific, financial, and graphing calculators for sale today can be programmed. Programs can add new and wondrous functionality to a calculator. The ability to program calculators can trace its ancestry back to the era of my HP-20S. In fact, one of the programs presented in this book is based on a similar program that I keyed into my HP--albeit, the program presented in these pages is far simpler and easier to use than its 1980s parent version.

Once again, like the previous smartphone debate, programming isn't the sole domain of calculators. You can program computers, too. Although it is pretty tough to argue against a handheld calculator running a unit conversion program versus a unwieldy computer performing the same task. Talk about overkill. You might as well try to insert two brains into one skull. Oops, I'm getting ahead of myself, again.

So choose your calculator very carefully. The model you purchase today could be your champion 30 years from now.

> Professor F. Stein, Emeritus
> Department of Electrode Stimulation
> University of Transylvania
> October 31, 2016

IN 1957, CASIO COMPUTER COMPANY LTD. LAUNCHED THE 14-A CALCULATOR WHICH USED 342 ELECTRIC RELAYS FOR SOLVING ADDITION, SUBTRACTION, MULTIPLICATION, AND DIVISION PROBLEMS UP TO 14 DIGITS.

INTRODUCTION

Greetings and felicitations. Thank you for purchasing this comic book. You are one the few really smart people in today's world. Why? Because, I'm guessing here, you own a calculator. Yup, where the number of total worldwide calculator owners ranked in the upper millions during the halcyon days of 1970 - 1980s (by some estimates almost 50% of the world's population), your breed has now shrunk to a teensy, if not appalling, estimated 300,000 - 500,000 people! Or, as expressed in calculator terms: 0.05% of 7.4 billion*, give-or-take a couple of hundred thousand.

*Hamrick, Kathy, 1996, "The History of the Hand-Held Electronic Calculator," *The American Mathematical Monthly*, Vol. 103, No. 8: pgs: 633-639.

I'll admit it, though, I have a soft spot in my heart for calculators--I love 'em. Being able to pickup a calculator and perform a quick number base conversion while programming an Arduino is very convenient for me. Likewise, calculating real estate mortgage rates, figuring price/ounce in the grocery store aisle, or determining miles/gallon fuel efficiency at the gas station are all child's play with my calculator. I would be loathe to entrust these utilitarian calculations to my luxury smartphone.

"A calculator in every pocket," should be a common rallying cry, but, alas, it isn't, ... yet.

What's wrong with this picture? Well, the ground swell acceptance of the smartphone has rendered the calculator as a "non-essential" device to most people. Unfortunately, they couldn't be more incorrect. Sure, these smart devices come equipped with a calculator, or, special mobile operating system Applications (Apps) can be added to them for providing an enhanced calculator-like operation. Unfortunately, all of this wonderful technology stops at most classroom doors. Furthermore, don't even think about taking a smartphone into a nationally-sanctioned testing center; no smart devices are allowed inside during testing. Although, to be totally honest here, sometimes calculators are banned from testing centers, too.

Luckily, the "big 4" calculator manufacturers, namely Casio, Hewlett-Packard, Sharp, and Texas Instruments, are making some significant inroads into classrooms and testing centers. Supported by budget-friendly initiatives, more and more school districts are finding calculator purchases to be more cost effective than pricey, high-maintenance smart devices--most calculators lack time-consuming software update and upgrade cycles. Likewise, many calculators have input/output (I/O) ports on them that enable direct connection to real-world sensors and testing apparatuses. Additionally, most of these sensors are already being used in classrooms, so this type of interface is a win-win for schools everywhere.

One slap to the face for using calculators that is very tough to ignore, however, has been the difficulty in enabling students to "learn" how to tap into the complete potential of the calculator. Yes, there's only and handful of keys, but generally, each key has multiple functions. Manufacturers have tried to lessen the steep learning curve associated with two-, three-, and four-function keys by employing math-like or textbook displays (i.e., displays that depict formulas just like they appear in mathematic textbooks), but students still end up saying, "Where's the equals sign?" Or, "How do I solve for X, when only Y is available?" This comic book is an attempt to help answer these and a lot of other questions about achieving the full potential of the modern calculator.

Granted, there are a lot of "how to" books about using calculators filling Earth's largest bookstore, but none of these feature Professor F. Stein! Nor, are there any comic books about using calculators. In other words, your currently holding a couple of unique publishing "firsts" in your hands right now. Congrats.

Enough talk already, let's get on to the comics. Or, as this cute little red-haired girl used to say to me, "See you in the funny pages!" While I think that might've been intended as an insult, this girl did become the current Mrs. Prochnow, so I'm not so sure, it might've been a prophecy, instead. I hope you enjoy the book.

LANDMARK TIME: AS OF DECEMBER 31, 2006, CASIO HAD SOLD ONE BILLION CALCULATORS WORLDWIDE.

A WORLD WITHOUT CALCULATORS; IT DOES NOT COMPUTE.

WITHOUT MATHEMATICS, THERE WOULD BE NO MUSIC.

WITHOUT MATHEMATICS, THERE WOULD BE NO SCIENCE.

WITHOUT MATHEMATICS, THERE WOULD BE NO MONEY.

WITHOUT MATHEMATICS, THERE WOULD BE NO TIME.

WITHOUT MATHEMATICS, THERE WOULD BE NO POINT.

GREAT CALCULATOR

STILL BAD CALCULATOR

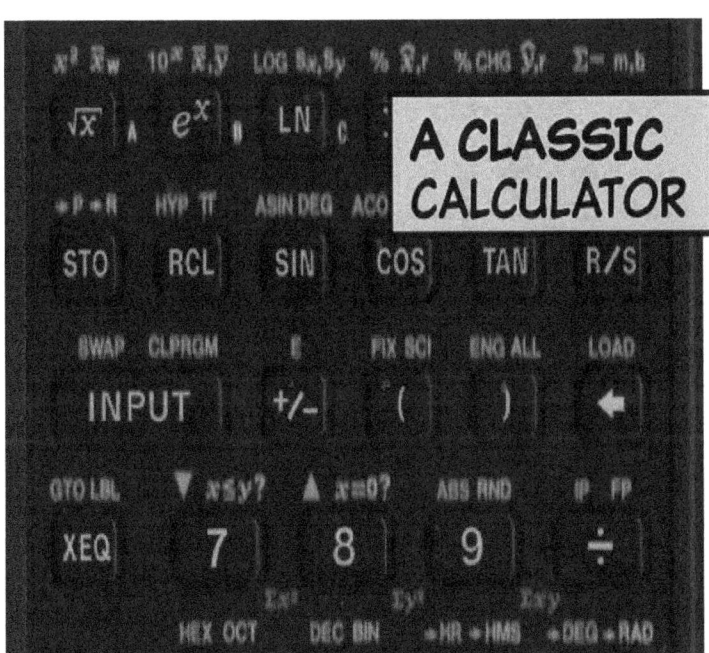

A CLASSIC CALCULATOR

NOW LISTEN, THIS IS A BAD CALCULATOR

STEER CLEAR OF IT,

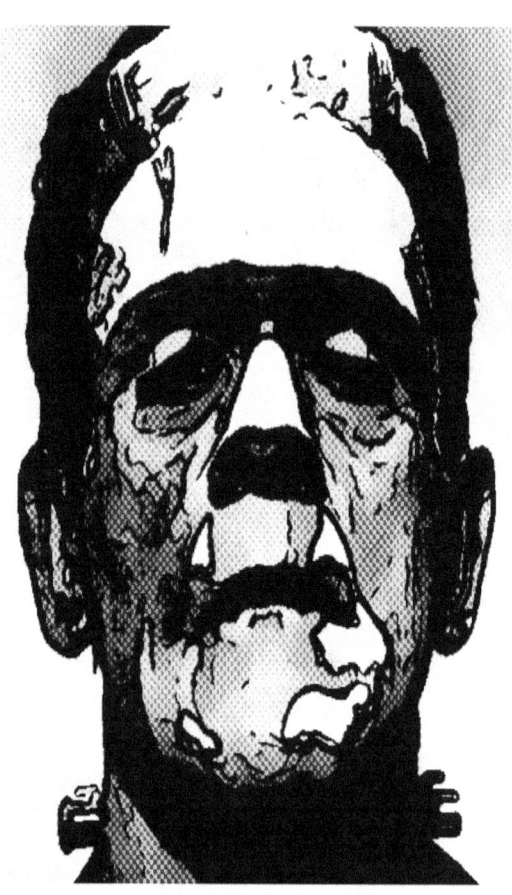

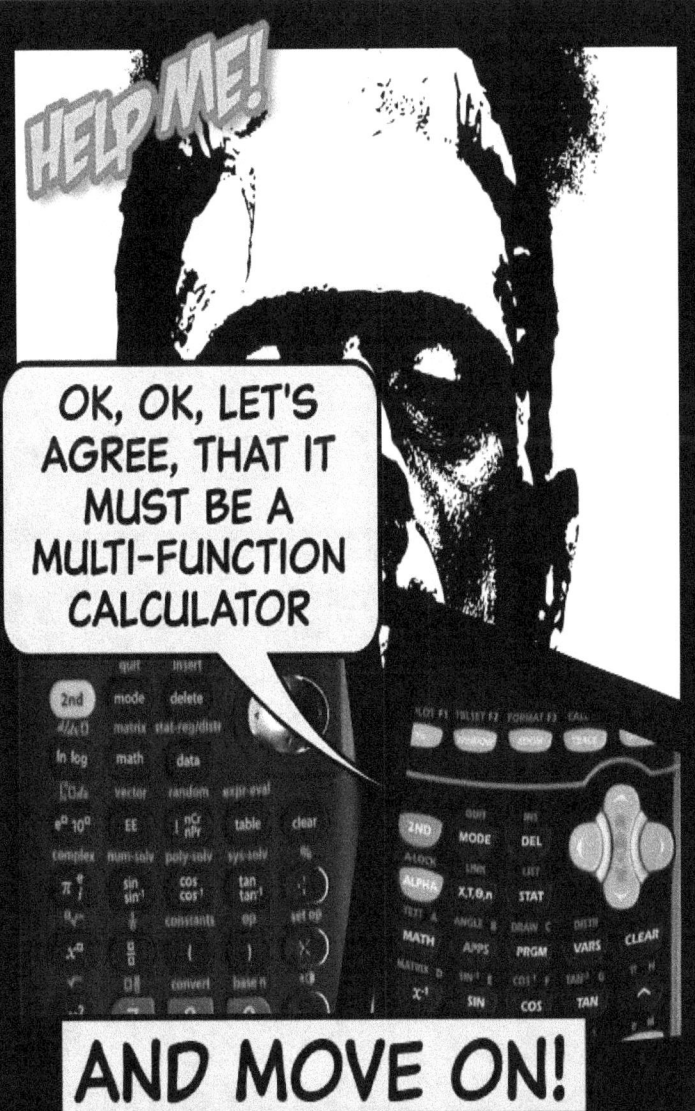

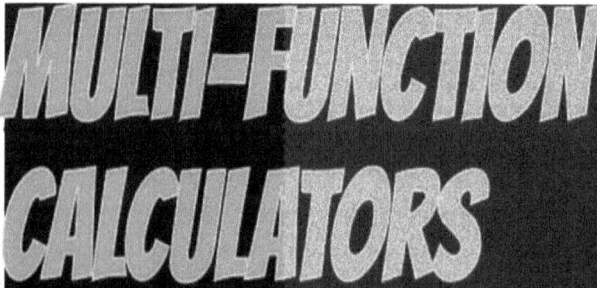

MULTI-FUNCTION CALCULATORS

TYPICALLY, MULTI-FUNCTION CALCULATORS HAVE AT LEAST ONE SPECIAL "SHIFT" KEY.

1 KEY

THESE SHIFTERS DISPLAY THE COLOR-CODED FUNCTION, ACTION, OR WORD THAT IS PRINTED ABOVE EACH KEY.

2 KEYS

HEWLETT-PACKARD INVENTED THE WORLD'S FIRST HANDHELD SCIENTIFIC CALCULATOR, THE HP-35 SCIENTIFIC CALCULATOR. THIS CALCULATOR WAS SO-NAMED BECAUSE IT HAD 35 KEYS. DEVELOPMENT WAS INITIATED IN 1971 BY BILL HEWLETT, HIMSELF. EVEN THOUGH IT WAS PRICED AT $395, THE HP-35 SOLD OVER 300,000 MODELS BEFORE IT WAS DISCONTINUED IN 1975.

HOME, HOME ON THE SCREEN

THE HOME SCREEN STARTS UP BLANK.

TYPE: [2] [+] [2] AND PRESS [ENTER] OR [SOLVE] OR [=].

```
2+2
           4
■
```

NOW STORE [2] IN THE VARIABLE T: [2] [STO] {ALPHA}[T].

```
2+2
           4
2→T
           2
```

FINALLY, TYPE: [2] [+] {ALPHA}[T] AND PRESS [ENTER]

```
2+T
           4
```

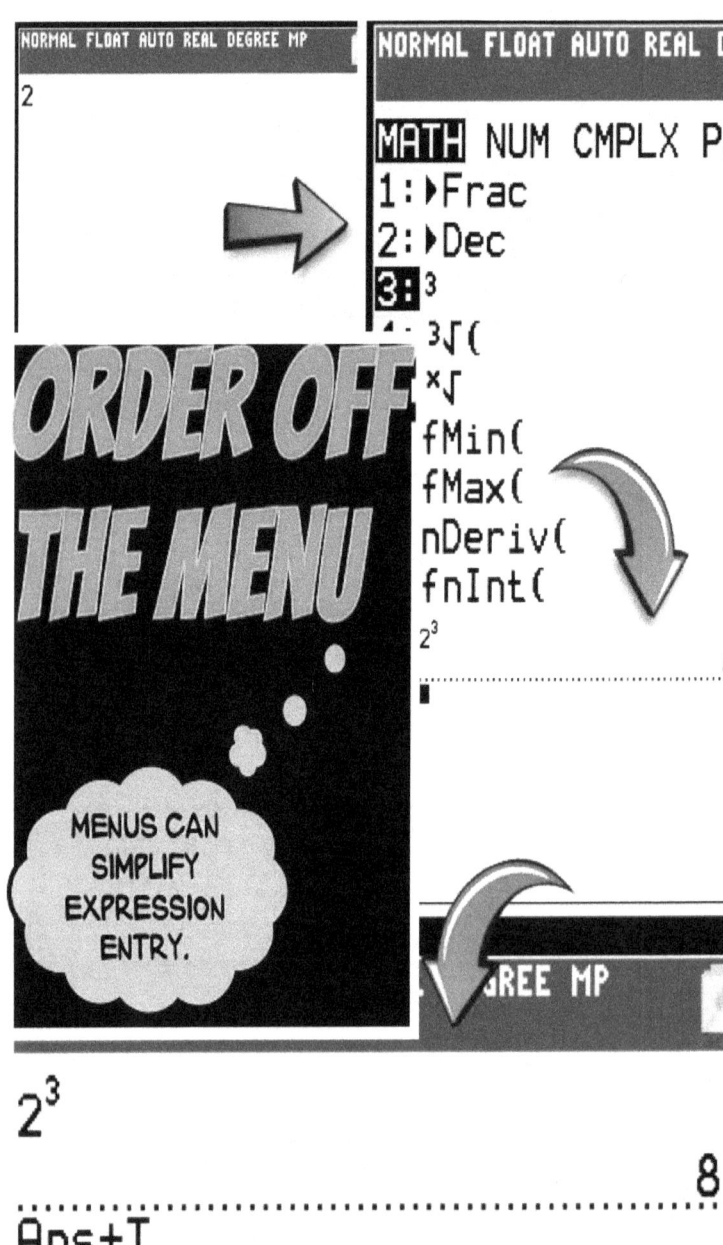

```
CATALOG
 LinRegTTest
 ∆List(
 List▶matr(
 ln(
 LnReg
▶log(
```

ORDER FROM THE CATALOG

ACCESS FUNCTIONS

```
NORMAL FLOAT AUTO REAL DEGREE MP

log(T)
                .3010299957
```

MAKE A LIST

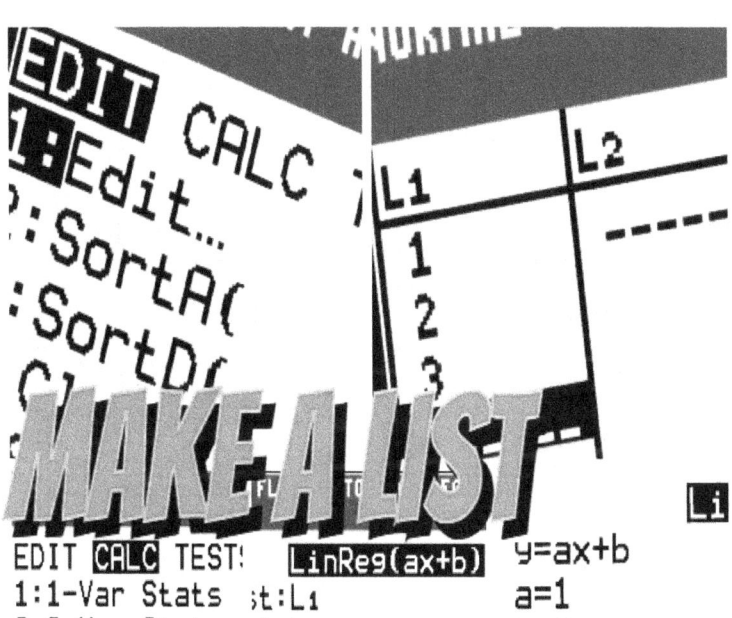

USE LINEAR REGRESSION ON 2 LISTS.

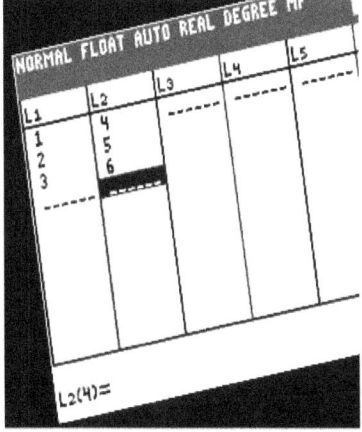

AREA OF A CIRCLE

WHAT IS THE AREA OF A CIRCLE 40 MM IN DIAMETER? | 20

1256.637061

HINT: PIE ARE ROUND, CORNBREAD ARE SQUARE.

```
NORMAL FIX1 AUTO REAL
DECIMAL SETTING
MATHPRINT CLASSIC
NORMAL SCI ENG
FLOAT 0 1 2 3 4 5 6 7 8 9
RADIAN DEGREE
FUNCTION PARAMETRIC POLAR SE
DOT-THICK THIN DOT-THIN
SEQUENTIAL SIMUL
REAL  a+bi  re^(θi)
FULL HORIZONTAL GRAPH-TABLE
FRACTION TYPE: n/d  Un/d
ANSWERS: AUTO DEC FRAC-APPROX
GO TO 2ND FORMAT GRAPH
STAT DIAGNOSTICS:
STAT WIZARDS: ON
```

```
20→R
                    20
πR²
           1256.637061
round(Ans,1)
                1256.6
```

YOU CAN SET DECIMAL ROUNDING SYSTEM-WIDE WITH FLOAT.

TAKE A NUMBER AT THE DMV*

*DEPARTMENT OF MATH VALUES

```
ANGLE              32°12'55"+10°6'13"
1:°                              42.3:
2:'
3:r
4:▶DMS
5:R▶Pr(
6:R▶Pθ(
7:P▶Rx(
8:P▶Ry(
```

ADD 32 DEGREES, 12 MINUTES, 55 SECONDS TO 10 DEGREES, 6 MINUTES, 13 SECONDS ROUNDED TO 2 DECIMAL PLACES.

```
ANGLE              32°12'55"+10°6'13"
1:°                              42.32
2:'                Ans▶DMS
3:r                          42°19'8"
4:▶DMS
5:R▶Pr(
6:R▶Pθ(
7:P▶Rx(
8:P▶Ry(
```

DISPLAY THE ANSWER IN DEGREES MINUTES SECONDS FORMAT.

POLAR COORDINATE EXPRESS

```
ANGLE
1:°
2:'
3:r
4:▶DMS
5:R▶Pr(
6:R▶Pθ(
7:P▶Rx(
8:P▶Ry(
```

```
R▶Pr(2,10
              10.2
R▶Pθ(2,10
              78.7
P▶Rx(2,10
              2.0
P▶Ry(2,10
               .3
■
```

```
R▶Pr(2,10
■
```

```
ANGLE
1:°
2:'
3:r
4:▶DMS
5:R▶Pr(
6:R▶Pθ(
7:P▶Rx(
8:P▶Ry(
```

```
R▶Pr(2,10
              10.2
R▶Pθ(2,10
              78.7
P▶Rx(2,10
              2.0
■
```

YOU DON'T HAVE TO USE CLOSING PARENS.

| log(4)■ | log(4) |
| | .6 |

CAPTAIN'S LOG

EZ LOGARITHM FUNCTIONS & EXPONENTS.

log(4)	
$\log(10^4)$	$\log(10^4)$
	4.0
ln(4)	
	1.4
■	

ZAAAAP

log(4)	
	.6
$\log(10^4)$	
	4.0
ln(4)	
	1.4
ln(4)/3	
	.5
■	

$\tan^{-1}(26/15)$

............................60.0

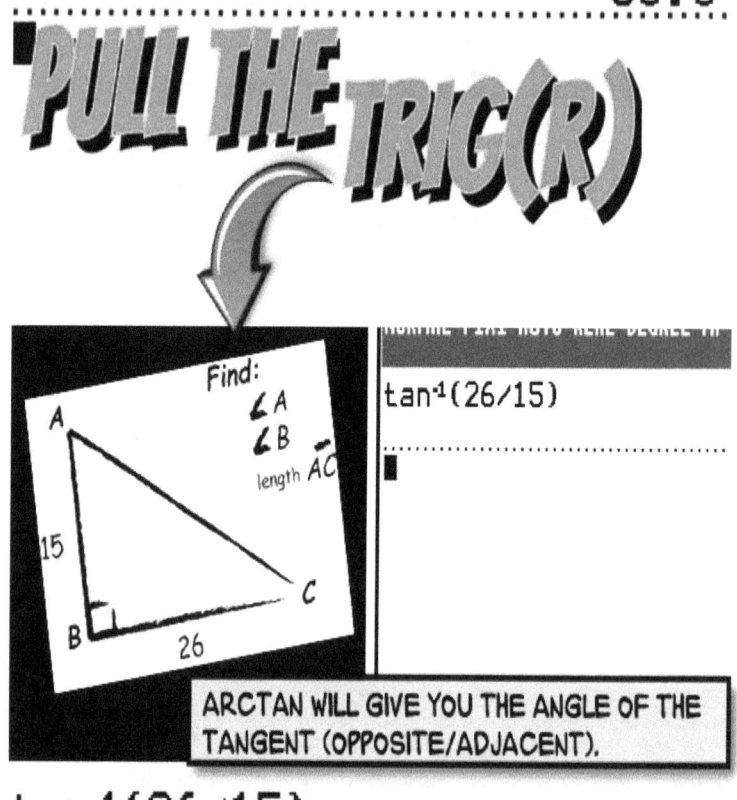

ARCTAN WILL GIVE YOU THE ANGLE OF THE TANGENT (OPPOSITE/ADJACENT).

$\tan^{-1}(26/15)$

............................60.0

90−Ans

............................30.0

$\sqrt{15^2+26^2}$

............................30.0

NORMAL FLOAT AUTO REAL DEGREE MP

HAVING TROUBLE FINDING THE = SIGN?

TEST LOGIC
1: =
2: ≠
3: >
4: ≥
5: <
6: ≤

EQUALITY FOR ALL EQUATIONS

TRY: [2ND] {TEST} [1] =, ERR, EQUALS.

LOOK FOR EQUALITY AS A "TEST" CASE EVALUATOR.

OTHER HP CALCULATOR FIRSTS INCLUDE: WORLD'S FIRST POCKET-SIZED BUSINESS CALCULATOR (HP-80 IN 1973); THE FIRST FULLY PROGRAMMABLE HANDHELD CALCULATOR (HP-65 IN 1974); THE FIRST HANDHELD WITH CONTINUOUS MEMORY (HP-25C IN 1976); AND THE FIRST CALCULATOR THAT DISPLAYED BOTH NUMBERS AND LETTERS (HP-41C IN 1979).

```
1:rand
2:nPr
3:nCr
4:!
5:randInt(
6:randNorm(
7:randBin(
8:randIntNoRep(
```

n:10
p:1/2
repetitions:10
Paste

randBin

FLIP A COIN

THERE IS A 50-50 CHANCE THAT FLIPPING A COIN WILL RESULT IN ONE OF TWO OUTCOMES: HEADS OR TAILS*.

*UNLESS YOU'RE IN "THE TWILIGHT ZONE" EPISODE, 'A PENNY FOR YOUR THOUGHTS' (1961), WHEN HECTOR B. POOLE FLIPS A COIN AND IT "LANDS ON ITS EDGE."

```
randBin(10,1/2,10)
        {5 4 4 8 5 4 5 4 4 5}
```

```
Bin(10,1/2,10)
    {5 4 4 8 5 4 5 4 4 5
Ans→L1
    {5 4 4 8 5 4 5 4 4 5
```

MORE FLIPS

THE RESULTS CAN BE GRAPHED; SHOWING THE DISTRIBUTION OF "HEADS" RETURNED BY THE BINOMIAL RANDOM NUMBER GENERATOR.

Plot1 Plot2 Plot3
On Off
Type:
Xlist: L1
Freq: 1
Color: GREEN

ZOOM MEMORY
1: ZBox
2: Zoom In
3: Zoom Out
4: ZDecimal
5: ZSquare
6: ZStandard

ADJUST WINDOW & ZOOM

Xmax=10
Xscl=1
Ymin=0
Ymax=10

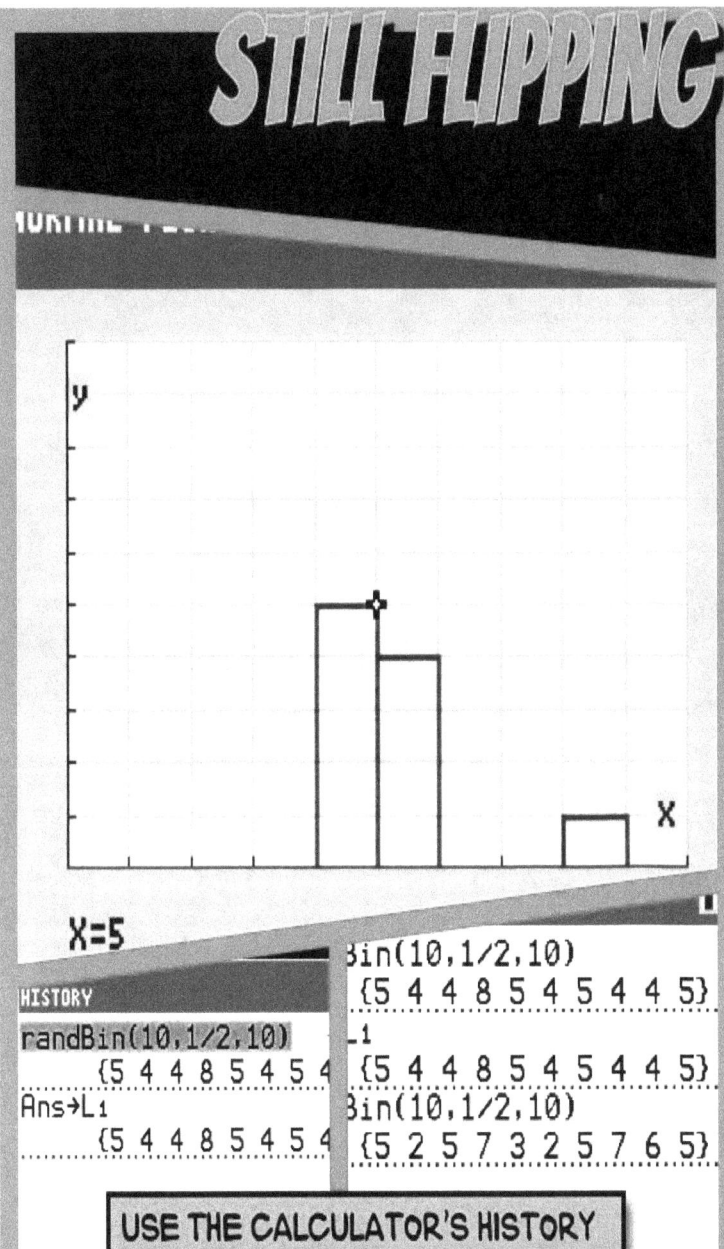

A DIFFERENT KIND OF FLIP

WHAT ARE YOUR CHANCES OF GETTING 10 QUESTIONS CORRECT ON A 20 QUESTION MULTIPLE-CHOICE TEST?

OH, AND YOU GUESS AN ANSWER ON EVERY QUESTION!

USE A BINOMIAL PROBABILITY DENSITY FUNCTION DISTRIBUTION TO FIND OUT.

```
      AUTO REAL DE    MP

DISTR DRAW
8↑X²cdf(
9:Fpdf(
0:Fcdf(
A:binompdf(
```

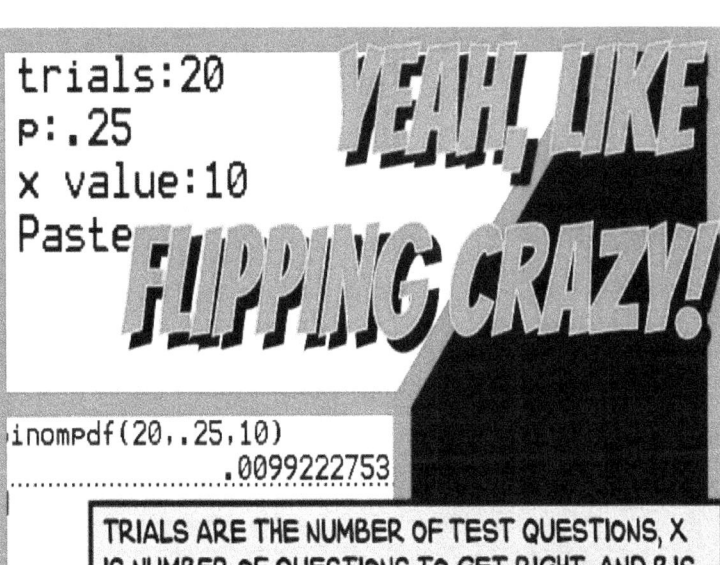

6: fMin(
7: fMax(
8: nDeriv(
9: fnInt(
0: summation Σ(
A: logBASE(
B: Solver...

RMAL FLOAT AUTO REA
TER EQUATION E1=E2

EQUATION

E1: $16X^2$

E2: 32

QUADRATIC PARABOLA (VERTEX)

VILLAGERS ARE STORMING THE UNIVERSITY'S DORM.

SO, YOU THROW A GINORMOUS BOULDER OUT YOUR WINDOW AT THE THRONG BELOW.

THE DORM WINDOW IS 32' HIGH. HOW LONG WILL IT TAKE THE ROCK TO SMASH INTO THOSE PESKY VILLAGERS?

USE THIS FORMULA:
HEIGHT=-16*TIME^2+32

OR,
Y=-16X^2+32

Y=HEIGHT OF GROUND
OR, 0
X=TIME TO HIT

HMM, SIMPLIFY IT:
16X^2=32

NORMAL
SELECT VARIABLE; PRESS

$16X^2=32$

X=1.4142135620...
bound={-1E99,

REAL DEGREE
VARIABLE; PRESS ALPHA SOLV

$16X^2=32$

X=√2■

bound={-1E99,1E99}

ADJUST THE GRAPH'S DIMENSIONS.

```
WINDOW
 Xmin=-3
 Xmax=3
 Xscl=1
 Ymin=0
 Ymax=32
 Yscl=8
 Xres=1
 ΔX=.02272727272727
 TraceStep=.0454545
```

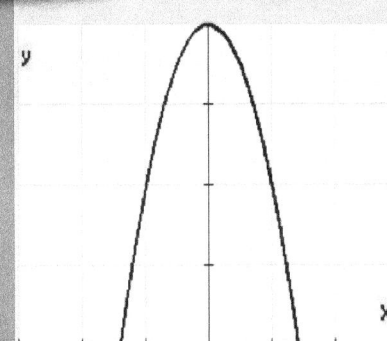

PLOT IT!

MOVE THE CURSOR FROM THE VERTEX (0,32) TO THE POINT ON THE X-AXIS WHERE Y=0.

IS IT CLOSE TO 1.414?

BINGO

LOL

QUADRATIC FUNCTION (STANDARD)

:Nederlan
:Periodic
:PlySmlt2
:Portug
:Prob Sim
:SciTools
:Svenska
:Transfrm

FLOAT AUTO REAL TRANSFORMATION GRAPHING

Plot1 Plot2 Plot3

\Y1 = 2X^2 − 4

IF YOUR CALCULATOR HAS APPS, LOAD THE TRANSFORMATION GRAPHING APP.

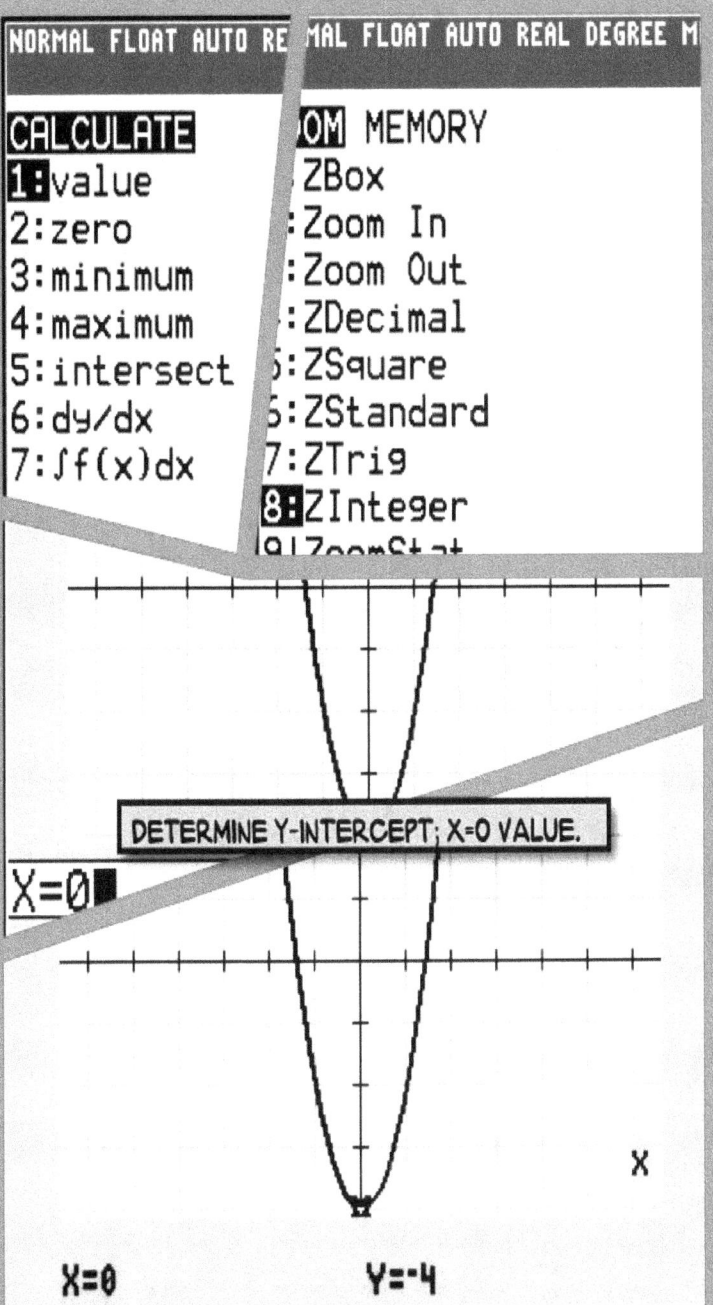

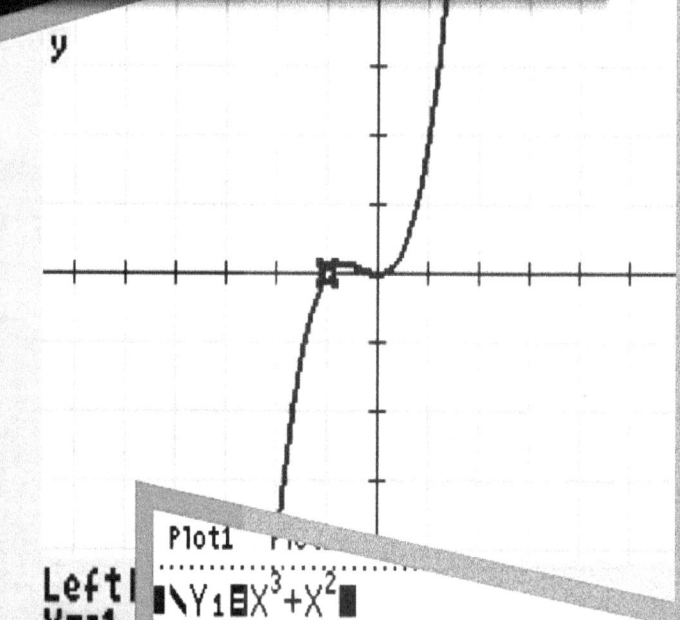

COMPLEX ROOTS

NORMAL FLOAT AUTO RE

Plot1   Plot2   Plot

\Y1=X²+9

SOLVE THIS POLYNOMIAL ALGEBRAICALLY.

\Y6=
\Y7=

NORMAL FLOAT A

$(\sqrt{-4*9})/2$

$3i$

$-(\sqrt{-4*9})/2$

$-3i$

MIXED ROOTS

$Y_1 = X^3 + X^2 + 4X + 4$

$Y_2 =$
$Y_3 =$
$Y_4 =$
$Y_5 =$

SOLVE THIS POLYNOMIAL BOTH GRAPHICALLY & ALGEBRAICALLY.

X = -1

$(\sqrt{-4*4})/2$
$\quad 2i$

$-(\sqrt{-4*4})/2$
$\quad -2i$

NATURAL LOG, STAR DATE E^X

SHOW GRAPHICALLY THAT THE INVERSE OF THE EXPONENTIAL (E) FUNCTION IS THE NATURAL LOG (LN) FUNCTION.

NOTICE THE NICE REFLECTION BETWEEN THE 2 FUNCTIONS.

```
Plot1  Plot2  Plot3
■\Y1 ▬ e^X
■\Y2 ▬ X
■\Y3 ▬ ln(X)
 \Y4=
■\Y5=
 \Y6=
 \Y7=
■\Y8=
```

```
Plot1  Plot2  Plot3
■\Y₁=10^X
 \Y₂=X
■\Y₃=log(X)■
 \Y₄=
■\Y₅=
 \Y₆=
```

BASE 10 OR COMMON LOG

WHEREAS, THE NATURAL LOG FUNCTION'S EXPONENTIAL REFLECTION IS BASE E, WHAT CAN YOU INFER ABOUT THE COMMON LOG FUNCTION'S REFLECTION?

GRAPH IT AND SEE.

$\log_2(16)$

4

$\log_2(8) + \log_2(2)$

4

KEEP THE BASE THE SAME.

4

$\log_2(2)$

4

$\log_2(4) + \log_2(4)$

4

A LARGE LOGARITHM'S VALUE IS EQUAL TO THE SUM OF ITS FACTORS.

GASP!

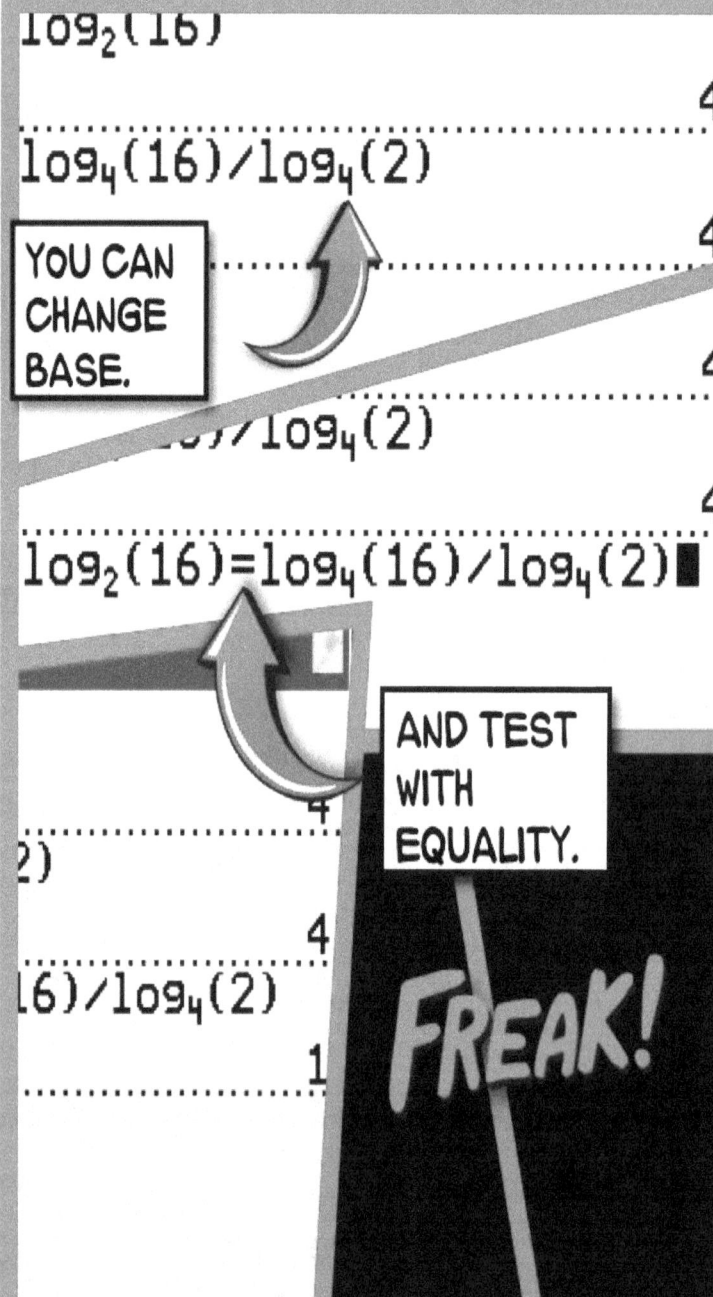

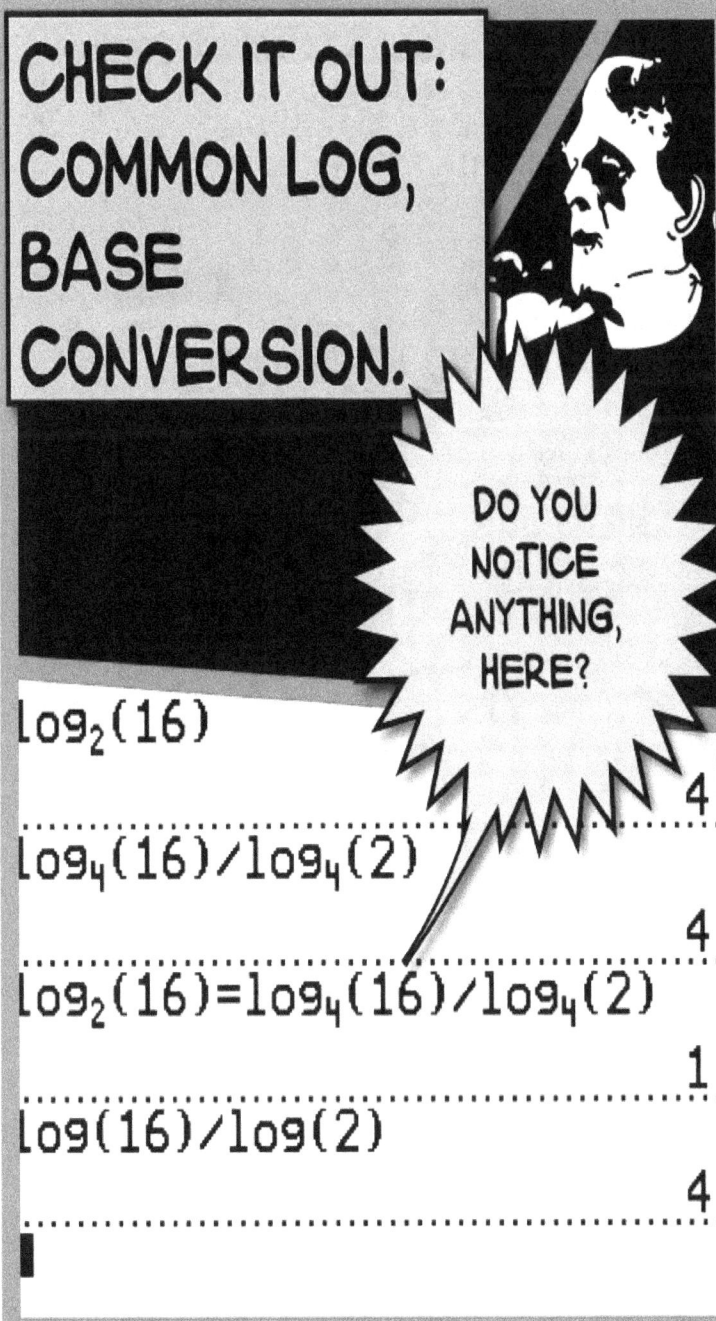

SO WHAT ARE LOGS GOOD FOR: OH, STUDYING THINGS LIKE SEASHELLS, ACOUSTICS, AND, EVEN, EARTHQUAKES...PLUS TONS OF OTHER NIFTY THINGS.

AND NATURAL LOG (LN).

$\log_4(16)$

4

$\log_2(16) = \log_4(16)/\log_4(2)$

1

$\log(16)/\log(2)$

4

$\ln(16)/\ln(2)$

4

YOU ARE NOW THE ACE OF BASE.

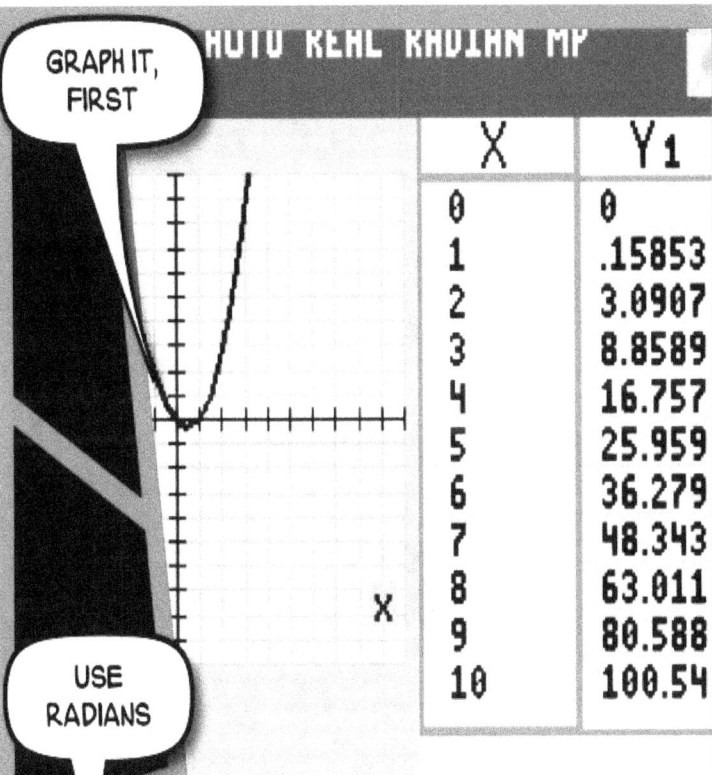

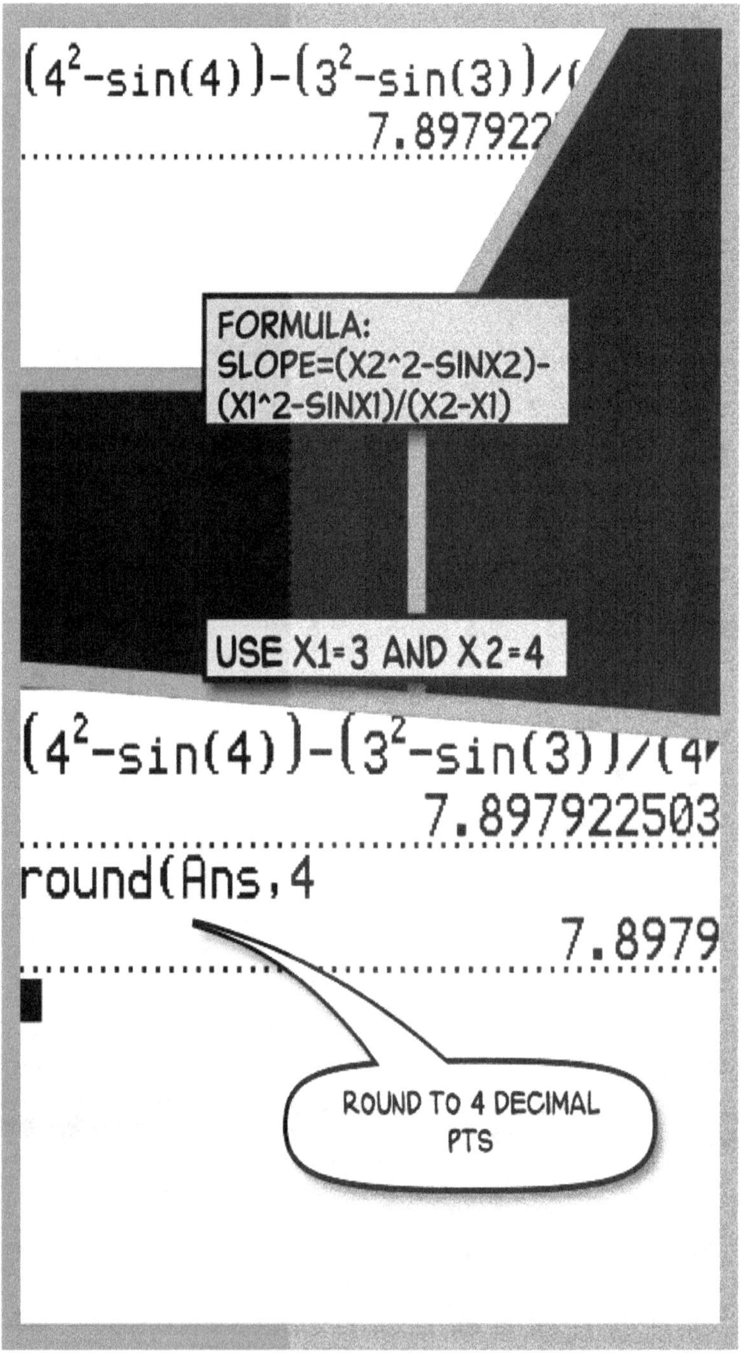

$(4^2-\sin(4))-(3^2-\sin(3))/($

7.897922

FORMULA:
SLOPE=(X2^2-SINX2)-(X1^2-SINX1)/(X2-X1)

USE X1=3 AND X2=4

$(4^2-\sin(4))-(3^2-\sin(3))/(4$

7.897922503

round(Ans,4

7.8979

ROUND TO 4 DECIMAL PTS

dayOfWk(
▶dbd(
DEC Answer
▶Dec
Degree
DelVar

AUTO REAL DEGREE MP

dbd(9.2216,12.2516

USE DBD: "DAYS BETWEEN DAYES"

CATALOG HELP

dbd(■

(date1,date2)

Programming Small, Medium, and Large

Small Programming

	$2 \to D$
USE THE [STO] KEY FUNCTION TO ASSIGN A VALUE TO A VARIABLE.	
NOW SUPPOSE YOUR BOSS WANTS TO KNOW THE CROSS-SECTIONAL AREA OF A 2" PVC PIPE?	$2 \to D$ $(D^2 * \pi)/4$ 3.1415
WRITE THE FORMULA USING YOUR STORED VARIABLE.	
OOPS, SHE MEANT 5" PIPE.	$2 \to D$ $(D^2 * \pi)/4$ 3.1415 $5 \to D$
[STO] THE NEW VALUE, RECALL THE FORMULA FROM HISTORY, AND	

```
2→D

(D²*π)/4
                3.14159265
5→D

(D²*π)/4
                19.6349540
```

YOU JUST GOT PROMOTED!!!

ROUND TO TWO DECIMAL PLACES

```
(D²*π)/4
                3.141592654
5→D
                         5
(D²*π)/4
                19.63495408
round(Ans,2
                     19.63
```

Medium Programming

You can learn all about using machine language (ASM) programming on TI calculators, here:

TIBASICDEV.WIKIDOT.COM

Beware some of these ASM hexcodes might not agree with your calculator's operating system and executing a faulty program could freeze your TI calculator. If this happens, try pressing the "Reset" button located on the back of the calculator.

Example:

Automatically display decimals as "textbook" fractions on the calculator's home screen.

1. Press [PRGM].

2. Arrow key to "NEW" and press [ENTER] for "1:Create New."

3. Enter a name for your program, like "HOMEFRAC" - Note: the max length for a name is 8 characters AND the first character must be a letter.

4. On the first line, search for your calculator's ASM Prgm function (e.g., Asm84CPrgm) using [CATALOG] and press [ENTER].

5. Make a new line by pressing [ENTER] and type this HEX code EXACTLY: FD360A0CC9 – Note: those are zeroes and NOT ohs.

```
NORMAL FLOAT AUTO REAL RADIAN MP

PROGRAM:HOMEFRAC
:Asm84CPrgm
:FD360A0CC9
```

6. Exit the programmer with [QUIT].

7. Execute this program with the Asm(function from [CATALOG] and the program name (see Step 3). You insert the program name by pressing [PRGM], arrow key to "EXEC," press [ENTER] for "1:HOMEFRAC," and press [ENTER] to execute the ASM program. If successful, you will see "Done" on the home screen.

Test your program by typing: .2 + .3.

```
NORMAL FLOAT AUTO REAL RADIAN MP

Asm(prgmHOMEFRAC
                                    Done
.2+.3
                                     1/2
```

Beware, using programs like this sample COULD alter mode settings on your calculator. For example, on a TI-84 Plus C Silver Edition, this sample program automatically switched from Real number mode to imaginary number mode. My only means for recovery was to Reset the calculator.

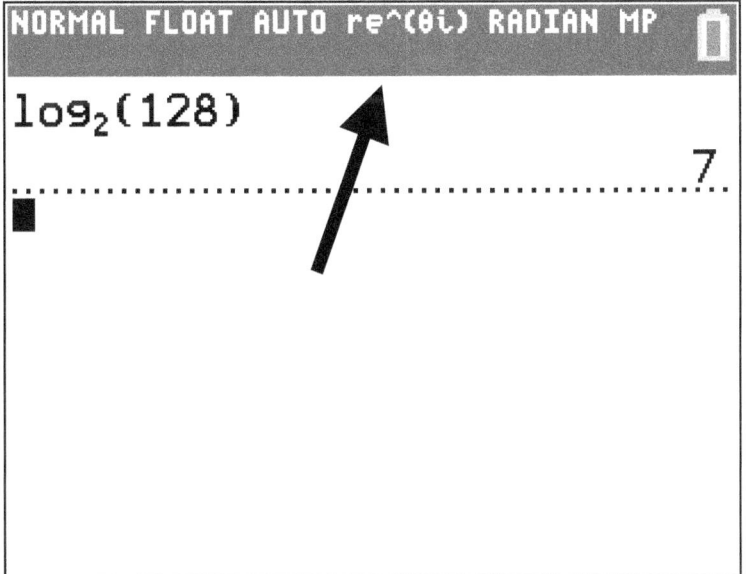

TEXAS INSTRUMENTS, INC., IN 1967, CREATED THE FIRST ELECTRONIC HANDHELD CALCULATOR.

Large Programming

Larger (and better behaved) programming for TI calculators can be performed with TI-Basic. Instructional resources for this language can be tough to locate--they are a Web-based "eGuide." Try this search pattern:

TI-84 Plus CE Coding eGuide > Coding References > TI-Basic Programming Guide for the TI-84 Plus CE Graphing Calculator, on the education.ti.com Web site.

You use the same [PRGM] key for entering your TI-Basic program into your calculator. Rather than using ASM hexcodes, however, you'll be using commands that are pretty easy to decipher. For example, "Input," "Prompt," and "Disp" are input/ouput (I/O) commands in TI-Basic.

```
NORMAL FLOAT AUTO REAL DEGREE MP

PROGRAM:DISCOUNT
:Prompt O,S
:(100*(O-S))/O→D
:Disp "DISCOUNT IS",round(
D,0),"PERCENT"■
```

In fact, an example TI program for calculating the discount of any sale item with a known original price and sale price would look like this when programmed with TI-Basic:

{name} PROGRAM:DISCOUNT
{command} Prompt O,S
{command} (100*(O-S))/O -> D
{command} Disp "DISCOUNT IS",round(D,0), "PERCENT"

Note: The little arrow symbol in the forumla is the [STO] key.

```
NORMAL FLOAT AUTO REAL DEGREE MP

EXEC EDIT NEW
1:DISCOUNT
```

```
NORMAL FLOAT AUTO REAL DEGREE MP

prgmDISCOUNT
O=?24.99
S=?12.48
DISCOUNT IS
                              50
PERCENT
                            Done
```

Just execute the program and input the orignal price (O) and sale price (S) when prompted, the result is the discount (D), expressed as a percentage. And one nice fringe benefit of this program is that it won't overwrite your calculator's modes, either. So now you can take your calculator shopping with you. And how cool is that?!?

Here's another TI-Basic program challenge that you can ponder:

Calculate the average speed by using this formula: AVGSPD = total distance/total time

How would you allow for the input of multiple distances and/or multiple times?

How would you compensate for different units of measure?

Simple challenges like this can turn you into a programming powerhouse. Good luck.

Show Your Sensitive Side with Sensors

Some calculators enable you to connect external sensors directly to them. Furthermore, this I/O connection is usually supported with special software that provides data collection and analysis right on the calculator without any support needed from a PC. These types of calculators then become powerful, rugged, portable data acquisition nodes for performing complex science and math activities where no PC dare (or, can) go.

Want to measure the pH of a local stream? Just grab a pH sensor-equipped calculator and off you go. Or, maybe you'd like to analyze the impact force of a boulder? Don't jeopardize your PC, use an inexpensive impact sensor-equipped calculator, instead.

One quick way for enabling sensor measurements with a calculator is the combination of a TI-84 Plus, TI-84 Plus C Silver Edition, TI-Npsire, or TI-Nspire CAS along with a Vernier EasyLink® USB interface (EZ-LINK). EasyLink plugs into the USB port on these calculators and automatically launches special data analysis software that is ready to obtain data from the attached sensor.

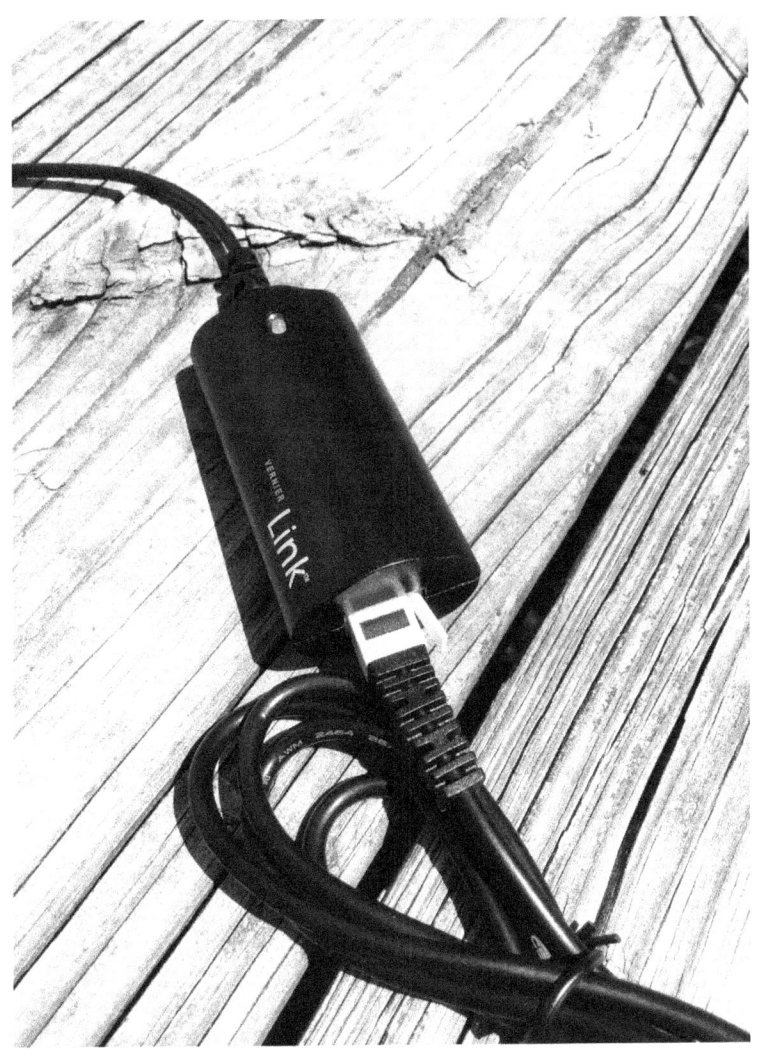

Don't have one of these calculators. Rather than using a calculator, Vernier sensors can also be connected to microcontrollers like Arduino. A special printed circuit board (PCB) manufactured by SparkFun Electronics acts as a simple "plug-and-go" interface between

Vernier sensors and Arduino. Unlike the calculators built-in software, however, you will have to program the Arduino for obtaining data from the sensors.

SparkFun Vernier Interface Shield - sparkfun.com/products/12858
Vernier Software & Technology - vernier.com

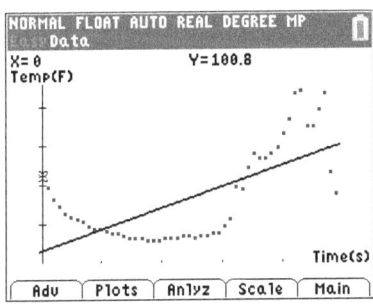

JUST A SHORT 9 YEARS PRIOR TO TI'S 1967 HANDHELD DEBUT, TEXAS INSTRUMENTS' LEGEND, JACK KILBY, SKETCHED A ROUGH DESIGN OF THE FIRST INTEGRATED CIRCUIT IN HIS NOTEBOOK ON JULY 24, 1958. IN A CAREER THAT INCLUDED MORE THAN 60 PATENTS, KILBY WON THE THE NOBEL PRIZE IN PHYSICS, THE NATIONAL MEDAL OF SCIENCE AND THE NATIONAL MEDAL OF TECHNOLOGY.

Testing, Testing, 0001, 0010, 0011

Determine a testing center's calculator policy BEFORE you go to take the test.

ACT Calculator Policy - act.org/content/act/en/products-and-services/the-act/taking-the-test/calculator-policy.html

PSAT/NMSQT or PSAT 10 Calculator Policy - collegereadiness.collegeboard.org/psat-nmsqt-psat-10/taking-the-tests/test-day-checklist/approved-calculators

SAT Calculator Policy - collegereadiness.collegeboard.org/sat-subject-tests/taking-the-test/calculator-policy

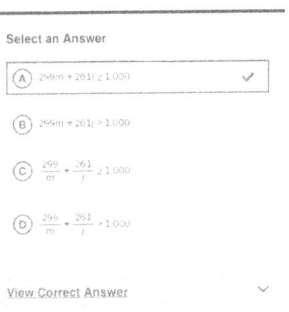

The first metacarpal bone is located in the wrist. The scatterplot below shows the relationship between the length of the first metacarpal bone and height for 9 people. The line of best fit is also shown.

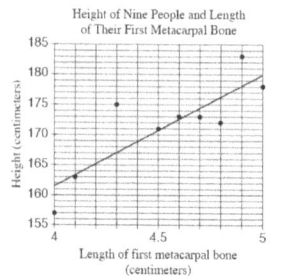

How many of the nine people have an actual height that differs by more than 3 centimeters from the height predicted by the line of best fit?

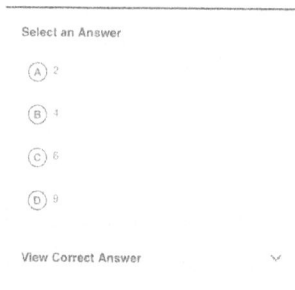

Select an Answer

(A) 2

(B) 4

(C) 6

(D) 9

View Correct Answer

Get a feel for taking each test with this suite of sample Qs.

ACT Practice Tests - act.org/content/act/en/products-and-services/the-act/test-preparation/math-practice-test-questions.html?page=0&chapter=0

PSAT/NMSQT and PSAT 10 Practice Tests - collegereadiness.collegeboard.org/psat-nmsqt-psat-10/practice

SAT Practice Tests - collegereadiness.collegeboard.org/sat-subject-tests/subjects?&affiliateId=aru%7Csubjtests&bannerId=

Calculator Magic Number Trick

1. Choose any three-digit number and enter it twice into the calculator. For example, choose "890" and enter "890890."

2. Divide this number by 11.

3. Divide the result from Step 2 by 13.

4. Now divide the answer from Step 3 by the original three-digit number from Step 1.

5. The answer is, ta-da, "7."

THE "C" ON THE TI-84 PLUS C CALCULATOR STANDS FOR "COLOR."

References

These Web sites can assist you with advancing your calculator knowledge:

Casio Resources – casioeducation.com/educators/

HP Educational Resources – www8.hp.com/us/en/prodserv/calculator/educators.html

SeeMathRun YouTube Channel – youtube.com/user/Seemathrun?ob=0&feature=results_main

Sharp Educations Support – sharp-world.com/contents/calculator/support/index.html

Texas Instruments Resources – education.ti.com

NOT ALL CALCULATORS HAVE THE ABILITY TO INSERT A PASSIVE PARENTHESIS AT THE CONCLUSION OF A FORMULA [E.G., SIN(X VERSUS SIN(X)]. SO WATCH OUT!

Formulas and Notes

Here are some formulas that you should know. Only a couple of them are printed here, the rest you can lookup on your own. Don't forget to add some illustrations, too. Good luck!

Distance between two points (x_1, y_1) and (x_2, y_2):
$\sqrt{(x_2 - x_1)^2 + (y_2 - y_1)^2}$

Sum of the interior angles for an n-sided polygon:
$180(n - 2)$

Number of diagonals for an n-sided polygon:

Area of a square = s^2

Area of a circle = πr^2

Area of a right triangle:

Area of a trapezoid:

Distance from a point (x_1, y_1) to a line $ax + by = c$

Angle between two lines:

Law of Sines:

Law of Cosines:

Volume of a sphere:

Volume of a cube: s^3

Volume of a circle:

Factor a polynomial:

Sum of an arithmetic sequence:

Sum of a geometric sequence:

Average speed:

Circumference of a circle:

Perimeter of a polygon:

Pythagorean Theorem:

Surface area of 3D polygon:

Statistics & Probability:

Standard deviation:

Mean, mode, median, average:

Derivatives: d/dx f(x)

Integrals: [a,b] f(x) dx

FINALLY

Here's an ASM program for TI-84 Plus C calculators that temporarily hides the cursor. It's great for running on that creepy kid's calculator that sits next to you in Math class. Oh, and the cursor will return when you turn the calculator OFF and then ON, again. Tee-he

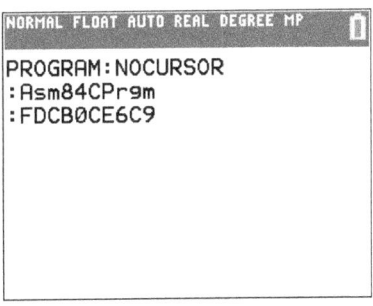

Colophon

This book was conceived, written, photographed, illustrated, edited, and published entirely on an Apple iPod touch*. The Apps used in the production of this book are:

Camera by Apple, Inc.
ComicBook! by 3DTOPO, Inc. (Jeshua Lacock & Joanna Mulder)
Pages by Apple, Inc.
Paper by FiftyThree, Inc.
PDF Expert by Readdle, Inc.
Photos by Apple, Inc.
Safari by Apple, Inc.

The fonts used for the text portions (not the cartoon pages) of this book:

3Dumb by Michael Tension
Architect's Daughter by Kimberly Geswein
Bangers by Vernon Adams
Komika-Text by Apostrophic Laboratories
Permanent Marker by Copyright 2011 Font Diner
SF-Cartoonist-Hand by ShyFonts Type Foundry

*Except for some calculator screen captures and font manipulation that were performed on an Apple MacBook Air and transferred to the iPod touch.

www.ingramcontent.com/pod-product-compliance
Lightning Source LLC
Chambersburg PA
CBHW060359190526
45169CB00002B/662